# THE AUTOMA

# THE AUTOMATIC ORACLE

Peter Porter

Oxford   New York   Melbourne
OXFORD UNIVERSITY PRESS
1987

Oxford University Press, Walton Street, Oxford OX2 6DP
Oxford New York Toronto
Delhi Bombay Calcutta Madras Karachi
Petaling Jaya Singapore Hong Kong Tokyo
Nairobi Dar es Salaam Cape Town
Melbourne Auckland
and associated companies in
Beirut Berlin Ibadan Nicosia

Oxford is a trade mark of Oxford University Press

British Library Cataloguing in Publication Data
Porter, Peter
The automatic oracle.—(Oxford poets).
I. Title
821    PR9619.3.P57
ISBN 0–19–282088–5

Library of Congress Cataloging in Publication Data
Porter, Peter.
The automatic oracle.
I. Title.
PR9619.3.P57A93  1987    821    87–14166
ISBN 0–19–282088–5 (pbk.)

Set by Rowland Phototypesetting Ltd.
Printed in Great Britain by
J. W. Arrowsmith Ltd., Bristol

*For Christine*

# ACKNOWLEDGEMENTS

ACKNOWLEDGEMENTS are due to the editors of the following periodicals in which some of these poems first appeared: *The Adelaide Review, The Age, Ambit, Encounter, Hubbub, London Magazine, London Review of Books, New Criterion, Poetry Australia, Poetry Book Society's Christmas Supplements, 1983, 1985, 1986, Poetry Chicago, Rialto, Scripsi, Slightly Soiled, Thames Poetry, Times Literary Supplement.* 'Throw the Book at Them' was one of the poems in *Invited Guests*, a collection of tributes to Kay Petrie (*Grand Piano Editions*). Some other poems were first broadcast on BBC Radio.

# CONTENTS

# A SOUR DECADE

These are the years which furnish no repentance
    Though seamed with sore regret:
    So much would selflessly be done and yet
Print no true sentence.

That grief sits down in books but is no writer
    Must be the just rebuke,
    And every lightless evening proves a fluke
The one grown brighter.

A careless management of things, they call it
    Who pose for God or Fate
    The purpose of the Infinite and Great
And here install it.

These decades, all the decimals of feeling,
    Are pressing on our schemes.
    On childhood walls, on corridors of dreams,
The paint is peeling.

# CLUTCHING AT CULTURE

That same purr-voiced disc jockey
has been too long among the toast today—
we've had them all,
a rave from the grave,
a rumble in the tumbril,
a suture on the future—
And breakfast used to be sanctuary
after the draught of dreams
and before the shop-front terror.
My daughter has gone out to work
and left me with the cat.
I look above my head
to the hardboard pinned with family snaps.
There she is, dressed in a tabard
with braided edges, holding tight the hand
of someone safe. A seraph
of the dangerous world, it seems
she's one step only from omnipotence,
as if she said with her unfailing smile,
Now is the ending of the world
and now goes on forever.

# RURITANIAN RESIDUALS

There, in the re-grown jungle,
a crashed Dakota or a Zero
and skeletons in khaki shorts
to be chanced upon by mineralogists—

Fathoms down, outside Murmansk,
after the acetylene intrusion
on the door-stopped ingots,
putty, cod-soft corpses in the dark—

Metaphors of human hope,
something we should not disturb
when we try the archaeology
of reminiscence—

At the Durbar of a thousand curry houses
the British Empire seems the greatest
piece of theatre the world has ever seen,
the gatling stuttering, the battleships in circles—

Thinking of love and duty, trying to hear
imagination's voice, the noise-floor
is too high, the Hooghly jammed with bodies,
a death-tide in the blood—

And drifting over, a cloud
of radioactive history presaging
cancer when? The spooks were always there
in Ruritanian redoubts—

If we executed one in seven
of our economists, the financial pages
would look rosier: but the curtain rises
on actors at an empty matinée—

Terrorists cannot stop material flow
or new Prep Schools appearing.
Listen in the Pleasure Gardens, that's
the Ghost Train running into Dachau—

3

The life of dividends, residuals
of the world's colossal comedies,
grosse Kleinkunst versus kleine Grosskunst,
nightmares of conviction—

And where to place the verbs in this?
Life has turned to pictures of us
as we were, and the Great Exhibition
plays host again to staring mendicants—

# THE PRINCE OF ANACHRONISMS

Adolescents smelling frangipanni
Watch with healthy hopelessness the girl
Next door whose breasts jog past the paling fence.
A Virgin by Matteo di Giovanni
Nets a million for a Seventh Earl:
Everything is in the present tense.

Screw those experts who insist they know
Tuscan economics or what Haydn
Heard when his french horns played B in alt.
Winds that live inside the mind may blow
Quarter on to quarter, they collide in
Safe Sargassos of imagined salt.

Friends, I love you but I will not buy
Transcendental global sophistry—
Nature's what it is, no shibboleth,
History's perspective is a lie.
Art is the same age eternally,
Births are anniversaries of death.

# STICKING TO THE TEXT

In the Great Book of Beginning we read
That the word was God and was with God,
And are betrayed by the tiniest seed
Of all the world's beginnings, to thrash
Like sprats in a bucket, caught in deed
As in essence by shapes of ourselves,
Our sounds the only bargains we may plead.

So starts this solipsistic essay about words,
Its first stanza chasing its own tail,
Since no word will betray another word
In this sodality, self-repressing and male,
And we discover, hardly believing our eyes
And ears, a sort of chromatic scale,
That whatever lives and feels is logos.

Tell us then, vanity, what is truth
And how does it differ from honesty?
Ecclesiastics and analysts play sleuth
To that slippery murderer, but they require
The rack and the couch, tell the story of Ruth
Out of her country, such cheats of championship
As the Noble Savage or General Booth.

We can know only what words may say
Though we may say what we know is untrue.
Honesty lines up its troops—Thersites,
Iago, Tartuffe, The Abbess of Crewe—
The confessional rolls, the lottery pays
Timely prizes to me and to you,
Truly honest people, tied to the wheel.

And when love announces it is here
Either with a lily and spasm of light
Or rising from a childhood bed of fear
To assume its pilgrimage of grace,
It brings its style wars and its gear,
The triolets of touch, the ribboned letters,
Pictures of Annus Mirabilis or just last year.

Keeping ahead of death and Deconstruction
We have the text we need to play the game,
But what should we do to make it personal—
Your text, my text—are they the same?
The rules are on the inside of the lid
As fate appoints its contestants and fame
Picks one from the Great House and one from the Pale.

Too many fortunes are made by the Absurd:
It's better to run in the linear race
Where everything connects which has occurred,
Better to suffer the nightmares natural to
The body and tell what you have heard
Among your fellow-sufferers and hope
The story's end won't choke you on a word.

Where do we go to live? We're born ticking
on the page and from the first disclosure on
we sense that time is useless without fear.
So here must gather all those claques of fact
we make good use of—and what are they
but words? Imagine the tight nucleus we know
is true inheritance: we find nothing more
to do with it than turn it back to chaos.
Proust could get ten thousand lines from
one night at a party and Robert Browning
knew he was in love only when he found he'd
said so on the page. How Elizabeth
loved his profile when it hovered over her
in trochees. Personification's special dangers
outweighed Daddy's growlings and the bladder
weakness of poor Flush. Rochefoucauld
spoiled things with his fully-frontal maxim:
it's all much cooler really, exile under cypresses
and chatting at the well, but never far
from the cherished self-immersing diaries—
no matter how fast they fill, white paper presses
on the eyes of nightmare and the black dog
barks defensively. There are mornings
in the bathroom when a wonky razor seems
*pons asinorum* of responsibility,
but don't despair, a brush with life's not final
till it's found a way to do the rope-trick
with dependent clauses. Dying's a book
with uncut pages; the pentel scurries and the tea
grows cold, and back in London a publisher
announces a burnished tome on Tuscany.
To get through life, just join the dots up, they
may prove a subcutaneous punctuation.
Today in Rouen there is an Avenue
Gustave Flaubert, but nothing spoils the stillness
at his desk. The DPP has all he needs
to start the trial—the boys in blue, the talkative
punk witnesses slurping from chipped cups.
The rules remain: you are the books you write.

# THE ACHING CATALYST

Flaking in a private room
All he has of fifty years
Hired from his mother's gloom
Seems a patronage of fears,
Balustrade and harbour-boom
More than gallows or death-bed,
Foolish gold not chosen lead.

Comfort finds a way to say
Greyness is a gloating light,
Abstract of the burning day
Changed to pictures in the night,
But the process slips away—
Here's the chair and he's here too
Guessing what the world might do.

World's the aching catalyst
Actions cannot do without.
Feelings think they can exist
Here and now and never doubt
Skill of eye and tongue and wrist,
But they camouflage the act,
Accessories before the fact.

*attacca*

the last Muse is waiting on the corner
in the coldest weather. There have been
arts and achievements piled as high
as ziggurats in a National Geographic
artist's impression of Babylon,
but real desire is a smaller thing
and bolder—it is to possess a body
just long enough for a part of ecstasy
to detach itself from the mainspring of the world
and come to you. You'll have time left over
to consider the Hanging Gardens . . .

ready to run after the public when
they've got ahead, saying 'Wait for me,
I renounce all my old obscurity'.
Temperament trips you up, you're left
with difficult books among the weekend shopping,
and still no grant. This is an old art
with too many new faces and the jargon
is always changing. 'Video killed the radio star'
as clouds drift down the valley, changing
green to grey to blue, assaults on misery
which must be hidden from your notebooks . . .

if those clouds went blue to green to grey
you would know the world had gifts, rain
upon a reservoir, the first drops speckling
its surface like the skin of trout and you'd say
grace for a blessing: the trouble is your shock,
too much talent in the world, it can't absorb
its own creation; there are queues in Heaven
as the million dreams fight to be born
and troop before the face of vindication—
these tribunes of themselves are pleased and helpless,
kissing a cheek held up in practised sorrow . . .

under this sign you will conquer,
go on and over the page to new worlds
and new movements. Run now down the hill
to a coast of high-cliffed islands
with a valance of bright picnic-grounds,
be reunited where they sell fresh chances
with a slimness long since squandered—
here must be forgiveness for existing,
'muddled and disappointing', and then she's *there*
your laconic Muse, bow-legged, in her wedding-suit,
instantly recognizable, coming up and saying . . .

# THE REST ON THE FLIGHT

The painted pale projectile moves and seems
Hardly to move to those inside its tide
Of twilight, as the close alliterative
Miles are ticked off charts of cloud and ocean
And still the pilgrim's face is held to sun
And still his stomach turns. A comb of words
Runs through his strands of thought continually,
Reminding him of God and lunch, of dust
At the end of day discountenancing shoes,
Of mango trees and winds in March. These props
Are human, more so than the glittering sprays
Of alcohol, the swimming-pools of in-
flight movies, or reminiscences of men
Returning to the Gulf. Packed into fear
With splendidly unlikely comrades, he
Out-herods Herod in this great escape
And relishes the massacre. Calling
The stewardess and asking for another
Miniature, he relaxes momentarily,
Is able even to risk formal scansion,
Seeming in this nowhere to stand up
Alone for seriousness, and yet frivol
In uncertainty—caparisoned
In blank verse in the blanker sky, his wings
Take him on to no place just to make
It plain to rulers of nativity
That godhead starts at home. Bring another
Whisky to the caravanserai—
At thirty thousand feet all solemn words
Are juggled out of sentence by the air,
What bubbles up is terror in its handy shape,
Heard through the headset as the ghost of hope—
Now you can write the history of the world,
Soon the little plates of Pentecost
Will be brought round accompanied by a tray
Of heated napkins, and the nose will dip
Towards the phosphor of a city; you're
Back once more to merchandising truth
Under wings of excellence, a miracle
Which Boeings make from immanence of sky.

# TO LACEDAEMON DID MY LAND EXTEND

The heavy spirit of ambition
tells you to write first this way,
then the next, to station envy
at the door of Nature's clubmates,
circulate among Time's franchise-takers,
simply to make the most of being here—
there is no coming back, don't sourpuss
the talent-network which is all you have.
                    Good advice, right feeling,
though it's well to think of standing
on a wood verandah, with the gale
flapping briefly through the canvas booms
and know just as the house lizards know
that legs of self are made for running.
                    Once life can sense its end,
the grinding ghost inside is what
the ego-soul most wants to let go of—
departed now the tenant who would suck his soup,
the housekeeper who moulded soap-ends
into parti-coloured balls,
fled with the breakfast post
the cashiered colonel who let you test
his regimental sword—
                    If these are fiction's butts
it makes no odds: life's particulars
are interchangeable at the end,
our memories written by the blood
in uproar. Judgement Day will be a raffle
or the spirit's hankerings
noted by the watchful staff—
the 'good' Russian table and the 'bad'.
                    Inside the egg
first stirrings are fatigued already:
Creation must create itself once more
to find the fire. When I was young
I drugged myself with opportunity
and so lay speechless on my bed all day
unplaiting shadows. Words which spoke to me
I locked in amber. Now I am by the water,
under vines and stars, waiting for nothing,

13

another glass of wine to hand, a page
cooled by turning: the inventory is short—
the ants, the kitchen light, a bat re-crossing,
a drying shirt, Doreen the Dog.

# A TRIBUTE TO MY ENEMIES

Apart from God I suppose I have none.
It takes some nerve to scale the heights
Of another person's privatest decorum,
I mean his hate. Better locate them,
Such hired riparians of straight reportage,
Jaundiced judges of a bootleg happiness,
In the sad intestines, losing out through age,
Opacity and alcohol.
                    But they would be useful,
Those enemies, to give my life a kind
Of focus. Not just besieging horrors
Scanned and sulked at in some North of England
Editorial, or subsidized experimenters
Glowering in grey-faced halls of education,
Hastening with terracotta folders to a class
Of willing Sufis—not even the towering
Yea-preachers whose laser scorn could track
Me down through all my dark of triteness
With the cry, 'You, especially you, don't matter.'
                    No, I must play creator
And make them up, these hierophants.
Invention is the mother of necessity
In the dreaming mind: fear abhors
A vacuum. Since God's on holiday
These are my shadows of impatience,
My nutrients of the future. Chiefs of Capability
Who can hold the corners of the world down,
Frogmen of the sexual swamp who have title
To all treasure, realtors, Realpolitikers,
Plainsmen with blue eyes come into town
Clutching translations of Catullus—
Would they be myself had I the courage?
                    I have dreamed them back so far
They are sexless, legless appetite in a mouth
Beneath a leaf waiting for the tell-tale shine
Of presence. This pad of words is just
My camouflage. I am the prototype
Of elderly animosity whose extenuation
Is to eat the words of love. I await them here,
My enemies in a thousand colours,
They will know me, I'm listening

To Bellini in the frosty dusk, the same
Performance once we heard at the Fenice
Vespering death. Music will last until
The brain has massacred its ambient creatures.

# HOME AND HOSED

*To whom it may concern,*
*the matter of disposal of my body at my death.*

I shall get the jokes over early.
My cat Flora is not likely to outlive me,
otherwise I would will her my feet and guts
and, of course, my genitals,
(about the weight of a small tin of Whiskas),
but we are still within the Unitary State
and may not feed our cats
on the riddling remnants of our lives.

No, consign it to the fire.
Take it to the undenominational ghats
set down drives of sycamores
beyond the golfing suburbs;
there the parsons are all locums,
prayers are culled from instances
of Anglican good taste and Presbyterian surprise,
the wreaths alone baroque,
and Bach without belief washes our formal faces.

Keep this shell in which I hermited
far from the black and invoice-laden men
who cleanse our world of death.
The priggish Baron who told Mozart's widow
not to waste her cash on coffins
wasn't wrong—'eine schöne Leich'
fits out the Viennese for festival
but Mozart's unmarked resting place
is inspiration to us all. I summon up
the silly cardboard castles and ideal lodges
of Australian Undertakers,
their presidencies of Rotary Clubs,
their Sunday programmes of soft music.
Surely the dead are livid at the living.

Hunter and hunted must come home.
Heaven which listened with ears of flesh
still sits insipid in the eaves of time:
Hell the forthright can't hold its own
against the newest movies; only blood,
blocked from tending to far-off oedemas,
races up-river to replete apocalypse.

I have limped far from high seriousness.
Assessors with my card in hand
will raise not a smile—'this flesh you were lent
is due for replacement, but yours is a case
without special merit. You made little sense
with your cries, bent no tonality
out of its rictus; you seem to us like
a sheep on an A road burst through a fence,
all tangle and terror. But, as proprietor
of some of our maker's immaculate filament,
you may call on the guarantee. Love,
great star of the terrified heavens,
shines on your head. It is time, true time
to give back your body. Know that the creek
keeps on flowing, the swallow skims to the wire,
someone is starting to climb up Saddleback.'

Soul's straightforwardness is crimped in the fire.

# PARADIS ARTIFICIEL

Barbel-cheeked and hammer-toed,
I'm scrambling up a river bank
in a landscape by Claude.

I'm not bothered by that group
of humid Muses, or Apollo's
booming cattle, his frogs with croup.

I am myself a metaphor
the painter didn't think of. I'm
the monster the readers are waiting for.

The countryside's classic,
stuffed with temples and nymphs,
glazed like boracic.

And into Arcadia, bump,
comes me. As Terry Southern said,
girls go for ghouls, hump! hump!

The music I make them hear
is genital intarsia,
kinaesthesia, or sex in the ear.

How did Europe get like this?
Look at the boring planning,
crepuscular taking the piss.

Out on a verandah the gods
are having a sundowner.
They live forever, poor sods.

Unlike the sharp-eyed fellow
who dreamed this picture up,
wetting his death-bed's pillow.

I'm off then, truffling through grass
like Nebuchadnezzar, all knees—
the artist can kiss my arse

His world's not a pretty sight,
he and his nymphs need me
to guide them through the night.

# PARADISE PARK

It is a time of the distancing of friends,
Of quarrels between the possible and impossible,
Of a face thrust at yours calling itself honest,
Of complex words outfacing polysyllables
Shouting through darkness to grey Menschenhass—
Now the sun is setting on the Great Divide
And there is nothing to do and nowhere to hide.


The New has been made and the afterwork sparkles
On barbicans of tenderized cement;
First Worlds and Third Worlds are equally fatigued
Importing and exporting barely describable butterflies
And alloys of the perfect miracle: trade goes on
Beyond Apocalypse and the dying and the hopeless,
Hurdles of misrule leapt lightly by the Press.


It is also a time of renunciation, of the artist
That master of paradox, becoming invaluable
By giving up hope, of his escaping his egotism
By abandoning his art. And who will notice,
Among the lengthening queues for Retrospectives,
A sacrifice so slight it moves no pivot's sliver
On warning systems in the nocturnal silver?


For, despite Doomwatch and its calibrated fright,
The human creature is forever coming into
Its inheritance, and out of little towns making
Famous hats and from stevedoring fastnesses
On unsavoury rivers, provincial hopefuls
Arrive at the capital, saving for the Weekend Ark,
The Violent Prater, the sails of Paradise Park.


You may sit at glass-topped tables with white wine
Or play the board games with commanding titles—
Conspiracy Theory, Life Support, The Pursuit of the Millennium—
But nobody gives up early, there is always a smile
Just behind the carousel's blind horse which hints
That being born is coming into pleasure and so who
Is waiting out there for you to be unfaithful to?

# CLEARING THE DESK

All night I rode his back, claws well in,
The cat of God endearing myself to
A crazed omnipotent, tearing his skin
And biting until his liver showed through—
      This is a dream, he said,
      It's only in my head.

It's come about through too much drink,
That and the acid rain of sleep.
Like eyeless Gloucester at the brink,
The Unconscious will not look, but leap.
      He kids himself, of course,
      A side-effect of remorse.

I get these assignments from time to time,
Self-accusers, all the self-regarding fools;
I am punishment, they must be crime
And dreams the multiple choice of schools—
      He is, no doubt, a liberal,
      But the age is mostly feral.

Astounding arbitrariness! I said in his ear,
'You must post the Holy Egg to my London Club.'
Explain that one, Dr Jung. In practice, fear
Is the pedal of evolutionary hubbub.
      What a song we all sing
      Going to meet the Demon King.

And now it's like a serial on the box,
Control has got wind of something big,
Our hero's sent to check certain stocks
And to bury the Egg beneath a Tuscan fig—
      No such luck, this is death,
      The real meaning underneath.

I've done my work for him. His desk is clear,
Letters willed to a National Library,
Manuscripts in folders, one for each year,
Incunabula collected. 'Stay with me,'
      He moans, 'the night is long.'
      And the cat's claws and the song.

# AND NO HELP CAME

Where would you look for blessing who are caught
In published acres of millennia
By ravishments of salt and raucous saints
Or janissaries drilling a Big Bang?
The parish of the poor you'd seek, far from
The high grandstands of words and notes and paints.

And when you drove your flagged and honking jeep
Among the huts of starving, brutalized
Dependents, you might chance to hear them playing
Sentimental songs of flowers and moons
Chiefly to keep them safe from art, whose gods
Build palaces adorned with scenes of flaying.

# TIPP-EX FOR THE OSCAR

Coming down from the high riparian park
whose municipal trees among the statued
victors of sad forts bear names of poets,
our this year's laureate finds some gristle
between two teeth and as he probes it free
he's struck by what will be preliminary
to the great excursion he is planning—
ah, now he has interior resources
for his invective; this Thalassotherapy
has worked. Among the salt-flecked palms he's found
his task laid out: an Ode to Western Ocean,
to that embarking which is a returning,
a nightpiece of the European soul.

Unpacking his Anglo-Teuton portmanteaux,
setting a timeloss on its true grief base,
he hears the grumpy oarstrokes round Odysseus
and Henry the Navigator's humpbacked ocean groan.
A child, he'd wondered at the starfish wheeling,
the men'o'war like lampshades, blue-bags
for his mother's washday—in fact the thousand 'likes'
expected by the listening crowd, and yet more subtle
metaphors, cartouches for those whose jaded
ears lack words and morbid eyes seek paraphrases;
he knew it would prove useful for the doors
of childhood to stand open in his lines
to observation, his seeming of the act.

But how to keep his poem in the fast lane?
Literature which feasts on poisoned tears
lies dead in libraries till the wind of fashion
bears it up: where will a living writer find
explicators who will say these tropes of sea
have led our estuarial history back
to our diurnal selves; we're made to mirror
islands in our allegories? You must have noticed
Kipling is rated in the University
Departments of English and that Shaw is not—
both famous in their time. O guide me God
and make me choose the words which flatter best
the passeggiata of the relevant.

23

The light stays on till 3 a.m. He's in
his room and fastened to his Ode. Downstairs
the smell of ham and cloves has not abated
its seaside boarding-house authentication,
but he can lock it out from here, his cave
of making. Sunk in his speluncan world
of Cabot, Camoes and even Kingsley,
the tempest acres of a great migration
surround his pen, tall fleets of hubris charge
across the Tropics, twilight men disturb
the Noble Savage and the *Monitor*
confronts the *Merrimac*. A thousand years
of history fill less than ninety lines.

Some faults (perhaps not faults but merely tics
of style) look out at him from what he's written—
tell-tale gruppetti in his adjectives
and clauses running into sand. But worse,
the pattern is not sticking, he can hear
more nervous coughs than claps inside the hall
of his imagination. Desperate for sleep
he seizes Tipp-Ex for those finished sheets
he'd loved to look on, and in fear he prays
the God of White-Outs to resolve his fate—
'O grant me those most blessed similitudes
which make from poets' words good novel titles.
The Ceremony of Immanence is gowned.'

# THE AUTOMATIC ORACLE

The nation knew that it was winning
By the applause of its diseases
And from application forms for brute longevity
Stuffed into every burgher's letter-box.
Many now took the Panglossian Grand Tour
On narrow-gauge circuitries past
Morganatic lakes, the landscape curiously like
A tamed Shantung, a province great
With cormorants and round-legged women washing.
Civil Servants changed to Public Servants
When the annual examinations grew
More and more parodic. Fancy coming from
A breakfast room of dogs-beneath-your-feet
To answer Question Ten: 'Complicate
The issues raised by Arms Control
Using only language of an Agatha Christie
Novel set in the Fertile Crescent.' Word changes
Were the windmills of our trade
And auguries of platitude. Why did everyone,
So long anterior to the Age of Steam,
Have such violent public cramps,
Exaggerate the anger of occasion,
For all the world like Cobbett on his horse
Dredging the public's whiskers for assent?
That time of prior seriousness was a dream,
A precise description of a nothing—
Travellers knew without enlightenment
That brothels backed up to cathedrals,
And the fields of rape were raised
For bored geographers. Colour me love,
A tone of weekends braving wading birds,
And in that hue I'll show a universe
Relaxed from taking thought. It did think once,
In a freshly furnished interregnum
Before fatigue invaded syllables,
But do not let such recollection spoil
Today's obsessive emptiness. It's our duty
To inspire our critics, high on history,
To new feats of rodent disapproval,
Reading, as they do, for decadent invention.
See how the tenses snaggle up the scene

In Modern Poetry: not making kingdoms
Of the murder in our minds we miss
An opulent and Tennysonian calm.
The world still hurts as much: it is the loss
Of register which boils the brains of each
Ambitious legislator. What was it,
Say pilgrims on the beach at Lerici
Trailing white toes through liver-coloured mud,
Gave that great pain its shape? Oh ask again
When the very trees have voices and the clocks
Drip death-watch on a home divinity.
Our age has its appropriateness, a playful
Sort of sepsis, and each of us is offered
Sponsorship of his own bookish end.
Meanwhile, mean temperature of elegance,
The state and its white sheets shall be
An oracle, one doing magic on
A minor scale. Walled by the sun
And burnished by the moon, our planet
Seems a garden from a catalogue.
Four feet, five feet from the mask of God
A veil is lifted and the sweet environment
Flowers like Easter Day. A world of innocence
Is back, but do not look too closely at
Its rich empleachment. Eden filters through
The public parks, the weekend wilderness,
As stiff as astroturf with fallen blood.

# LEGS ON WHEELS

Behold the Master Species
which haruspicates its faeces
and builds amazing churches
where an invisible God has perches,
a sort of playful demon
that christens itself human,
the top of the great tree
of evolutionary
existence. It lays out gardens,
uses up hydrocarbons,
has forgotten how to walk
but stuffs itself with talk,
jumps into the car
if it has to go a yard
to get the cigarettes
and hoards like nuts regrets.
A challenge to the seasons,
it has its special reasons
for poisoning itself
with recipes for health,
but though it screws the planet
from Andaman to Thanet
it dreams a Green Revival
and sponsors the survival
of all attractive creatures
(O hide your ugly features,
you vultures, snakes and rats
and purr you pussy cats!)
This arbitrary biped
invents a far-off Sky-Bed
where lies a scolding Father
working himself in a lather,
and having stripped the jungles
our very Prince of Mongrels
likes to settle down
in some well-hemstitched town
dotted with aforementioned churches
and rook-inviting birches:
there its family chariot
will so succinctly carry it

the sure suburban round
it hardly touches ground.
Light verse's inclination
is not to indignation
but yet its rhymes and rhythms
may choose the noble Houyhnhnms
above the Yahoos' shouting
on any public outing,
and then the doggerel canon
finds moral tales to hang on,
looks at stumpy legs
and foreheads tall as eggs,
applauds the Life Force which
inhabits touch and itch
and spurs us to the station
of superior adaptation.
May we avoid disaster,
escaping ever faster
from all the patent deaths
(Aids, cancer, madness, meths),
rejoice we swirl on wheels
instead of our bare heels
and changing down for climbing
perfect our human timing,
unnatural in our daring,
quite naturally despairing
of just our own devices
but clever in a crisis
show evolution that
we are the where it's at.

# WHO NEEDS IT?

Sitting the bad summer through
Above the square's leaf-thickened view,
    A writer might take heart
    From all that mars his art.

Though this is Nineteen Eighty-five
With Mankind strangely still alive,
    Selfishness asserts
    That more than his deserts

Shall be his luck, as word by word,
He battens on what has occurred,
    And makes some flimsy things
    Out of his happenings.

Thus, fours and threes, in Marvell's way,
May help him dangerously essay
    Horatian persiflage
    In a Martian Age.

For all the Caves of Making are
Solitary Confinement in a star
    And as for relevance,
    That's a mask of chance.

We can't write well unless we think
Of something unapproached by ink
    And yet we're caught in rules
    Of Writing/Reading schools.

It happens here, at the fibre's end,
The blood-sharp surge of words that send
    Codes of memory
    To the cortex-tree.

It shakes the branch of life, the leaves
Are strewn, and there between two thieves
    God is crucified—
    What if the story lied?

When the truth is known and told
The writer's face looks blandly bold
     But not one Jew is saved
     And Mankind's still enslaved.


Tremendous Art! But that's the real
Whirlwind which the people feel:
     In notes and paper, it
     Loses life a bit.


The poets, like the gods at lunch,
Are such a greedy fumbling bunch—
     Was beauty massacred here?
     They saw mascara smeared.


Things not for changing still cry out
And truth may even pack some clout,
     But artistic glory
     Is another story.


Our human genius—who feeds it?
And professorial art—who needs it?
     Stravinsky works all night,
     The others shouldn't write.


Our vanity may save us yet,
Dogs *do* get better—ask the Vet.
     Milton was sent to bat in
     Poor light, and *he* knew latin.


No reprimand will stop the writers
Proving fops and clowns and biters,
     But when they've gone, a line
     May fill with light and shine.


A toast then! Old Equivocators,
Envious friends, resourceful haters,
     All the way from Blimps
     To sub-suburban wimps.

Now praise the publishers, our kin,
Who do such good and don't get thin—
    Endure tautologies
    In famed anthologies.


The Pen is mightier than the Sword.
The Sword and Pen are in accord.
    Their power leaves the hand
    And decimates the land.


Thus each connives in dying light
To freeze the human heart in spite
    Of centuries of hope
    And Shakespeare, Donne and Pope.

# NEVERTHELESS

*Heretofore*
you could use words like heretofore
without embarrassment, and catch the tail
of lyricism in some suburban garden
where words were still silver
and stoically inventive verse stayed new
through all the limits of recurring sound,
being at once proconsular
and steeped in sadness. This was a time
when needles, once a shining store,
clicked in the dusk of cottage madness
under a maidenly predestined order
like the village poplars: a sad connective
linked complicity and fate,
Calvinism and the empty grate.


*Moreover*
what was more emphatic tipped over
into rhetoric, itself a form from which
the cooler mind could distance its approval
if the sought exaggeration seemed
too smug: viz. the walls of oak round England
and bloodless the untrodden snow.
Words went in carriages and knew
which world was which and who was who.

*Notwithstanding*
such non-literary facts as one may note
with understanding when the scientists
of history and language point to verse
as the least apposite of checks on how
the language went in any chosen past:
look to the lawyers and the letter-writers,
the estate agents in their coaches,
ironfounders and top-hatted railway men.
Now all the practical romantics
called from high-walled gardens
to God in a lunette: retreat was sounded
in that melancholy long withdrawing roar.
Circumlocution for these times
was the shape of Empire and an office
where words were banked or shovelled into coffers.


*Nevertheless*
we never move too far from Presbyterian
small print—wireless becomes the radio,
the word processor is a crutch for brains,
glossy paper takes the images of famine.
Poetry goes on being made from sounds
and syntax though even its friends confess
the sad old thing is superannuated.
Watch, though, its spritely gait as
like a bent mosquito in it walks
on the arms of three tall women in black
to mutter envious obscenities at large—
its task is still to point incredulously
at death, a child who won't be silenced,
among the shattered images to hear
what the salt hay whispers to tide's change,
dull in the dark, to climb to bed
with all the dross of time inside its head.

# ALAS, FLEETING

Oh, the Old Roman,
churning up the mud of our hearts
(What do you mean, old ?—
he was a brisk young placemaker
when he crammed his nose into his odes)—
Oh, the man we'd like to be,
choosing badly and recovering,
finding friends among the Villa Men,
those parvenus who are the only lords
to share the artist's sense of jeopardy,
the man who mixed regret and duty
in the spiciest yet blackest numbers,
who consoled untidy schoolboys
once they had become good District Officers,
talking all the time of how time passes,
how youth becomes cold useless age
and beauty drops away from limbs
like hoar frost on the apple trees
of Daddy's witless orchard—talking indeed
of passions gone which truth admits
were never there at all: Suave Moderate,
the wonder-working craftsmen trusting only
language and its un-pin-downableness,
patron saint of everyone born old
or timid. And the glimpsing sensuality,
a flicker of white lust, or like one horn
of a snail's reconnaissance,
the morning boys still sleepy in the showers,
a girl's arm trailed along the water's edge,
punting out of lilies. Wonderfully exploitable,
the human race! Not just time's erosion
but the spirit of all awfulness in the past
which willingly we won't remember, so that
hope comes round again and we are tricked.
My old headmaster, whose phrase this was,
following his entrance like a Disney dinosaur,
would purge the class: 'Eheu, fugaces,
can this be indeed the erudite Sixth?'
Horace would have hated him but known
how to placate him. At the end,

placating doesn't work, and yet he had
some odes stuffed in his pocket when he got there
and the bees of the underworld were singing,
'Welcome, poet, to our half-mast world,
Who'll do, does;
Who can, cans.'

# ESSAY ON CLOUDS

A complacent Gulliver, I lie
in silent dripping Norfolk
watching these flying islands .
with selfish unconcern—
here are planetary worlds
of silvered science
but I care only that they
block the sun from signing my dull skin.

So far down, this temperate garden
and such reefs above!
Oceans floating over us
and still we breathe the neophyte
scent of disengaging pollen.
The rug, my books, the cruising cat
are drowned with me,
we do not even seem to sleep
in our afternoon pavilion.

The clouds address me:
'you will never see us after this,
though our obliging cousins
will bring continuity,
but we have marked you, flying over,
the last one of the dynasty of self.'
I can calmly wait
for such archaeologists to find me.

Why is the sea in the air?
It's only books which say it is the sea,
the clouds abhor redundancy.
Now a black stripe and then
a pall of grey bring in
tormenting voices.
They are the sensitive ones
whose ears can hear the million cries
of animals in abattoirs.
The garden is sticky with their blood.

The sun comes out
to purge exaggeration.
The sun enjoys short sentences
but clouds prefer
a shifting Jamesian syntax.

Tea is brought out on the terrace.
Once more the clouds reproach me:
'because you are so incomplete
you cannot think of us without
dragging in yourself. You are fit
for nothing better than for prophecy.'

I watch one cloud come visiting.
In half an hour it disappears
to keep an appointment in the Wash.
I wave goodbye knowing I shall miss it
less than the passing cyclist on the road.

Night awaits the upper wind.
I decide I should not like to live
in a universe kept up by love
yet unequipped to tell a joke
or contemplate the sources of its fear.

# ESSAY ON DREAMS

To set the impossible as homework!
The only thing more boring than
someone else's dream is being told
the plots of films. So, to be arbitrary,
dreams are like Italian landscapes.

The known and felt and the surprising
brought together on a patterned plain
approached through car lots and untidy vines,
the Val di Chiana with a fattoria in sight
and irrigation jets spasmic in the grain.
You could be a caterpillar on a stalk
or Totila down on one knee
among the frescoes. A country you know is yours
but totally indifferent to you
where you cannot speak the language
and frogs leap into your hands,
sand underfoot like saintliness,
the modern and the ruined perfectly combined
until you're purged of every scent of home,
a true adventurer with the aegis,
warts appearing on your flying hands.

Gods cannot hide themselves. You see the spear
and brow of white Athena and you say,
What is the meaning of this masquerade,
you are not my mother. But others hide.
Training for death is what we do:
such ponderous mixtures, classic shrubbery
and a red poinsettia climbing to a door
of whitewashed lattice, heavyweight hubris
in two hemispheres . . . but try to make report
to any ghostly father and the dream will laugh
and settle in the curtains like a scream.

Look to the island where the herd of goats
is from a shipwreck and every dangerous
bird is widowed: seek to explain to the whole
houseparty that you'll reveal the murderer
tonight at cocktail time; find the faces
you were sure of sweaty now and snarling

as some tourists burst into the kitchen
accusing you of being Anti-American.
Arbitrary is arbitration
as the light inside the mind reveals
each dusty truth on its collapse,
and dreams are no one's coded messages,
merely the second life our flesh is fed.

Daylight drags itself through windows.
They are coming round collecting papers,
you must hand yours in. You have covered it
with nonsense, or left it white with fear.
The great poems you could write
are all assigned to Dickinson and Donne.
Peculiar to hear 'a person so obtuse
he could not even dream us a straight line'.

Round the promontory, bursting like a cake,
the ferry comes and soon its wave runs up
the slimy ramp. Perhaps this is the spot
where everything began; it could be we
can leave the toys out on the lawn and start
for home. Why not repeat yourself, the noisy
insects in the trees assert—all good is
habit that invades the homeless mind.
And dreams have never heard of history
or style, but like our childhood games
they knock us into love with present fear.

# A BAG OF PRESSMINTS

In the middle of a difficult book,
called I see 'The Allegy of Love',
I ask through a cloud of interference
or is it the Claud of Unknowing
why the spectral voice of truth
is fiddling with my chosen words.


It's not that these are simple messages,
God getting in touch with Pharaoh
or the Dog of Death mooching with the Furies,
rather it's the sky unscrambling
in its joy of chaos: everything in place
but huge disorder in our view of it.
The world was misprinted its first day.
In the beginning was the lord;
Coercion or Correction was His name;
He hungered in the Book of Genocide
and the great beast, the Adversary,
was Grauniad, the ever-eating mother,
appearing soon afterwards at the breakfast table
as reason's voice, The Guardian.


This is to snatch significance
from decadence of purpose. Language,
though it says it knows its place,
will always try to be subversive,
telling the eight a.m. analysand
that jokes or tumblersfull of sherry
won't blunten Grandma's teeth one bit
and little girls will never get out of the wood
even if they do know how to spell.


Here it comes again, like an ellipse
of the sun. I'm correcting proofs
in a dream and find I've written
'Patriots always stand for the
National Anathema.' And two lines later,
'Seven Hypes of Ambiguity'—
I'm quite relieved my misprints

make such sense. A pity my books
are kept so plain by too much meaning.

Help yourself to a strong mint from my bag,
it will take responsibility off your breath.

# RADIO CALIBAN

This is Imagination's nuclear-free zone,
so answer, airwaves, answer!
And hello to the girl
who asked for Ariel's new single.

Personally I'd take Wallace Stevens
to Prospero any day, and here's
that smouldering oldie, 'The Bermudas'—
Andy Marvell in the English Boat.

Time now for our popular feature
'My Most Wimpish Moment',
but first let's join Trink and Steff
singing 'I Cried to Dream Again.'

Our Studio Doctor says
women who eat meat grow more body hair—
Remember you heard it first on Radio Caliban,
the voice from the middle of the sandwich.

I saw God in an oleander bush,
writes Mrs Sycorax of Alfred Avenue
and, yes, Miss Mandeville, by starlight
a word may seem a planet. Hang on in.

Don Alonso of Bellosguardo
wants to know what became of
'The Thousand Twangling Instruments'—
same group, Don, I've got them on the brain.

If you don't want worms, lay off the cheese.
When the big wedding's on—Miranda
and Ferdy coming down the aisle—
all eyes are on the bridesmaids' boobs.

Watch an ant drag ten times its weight,
that's your blood holding back your death.
What you write makes sense or else words
would curl up in your palm like a paper fish.

Not much rhythm, not much art—
all right, but it's got feelings, listeners!
So let a lady wrangle with the blues,
'They flee from me that sometime did me seek.'

Absolute Milan where wise savers go.
A great video, those cloud-capped towers,
and Giorgione storming on the shelf—
a coral island ringed by sun and surf.

# TRY A TRIOLET

### 1

There are no seasons in this flat,
though out of doors the sun shines bright.

The open window tempts the cat.
There are no seasons in this flat.

Heat steps inside, the night comes pat.
Our dreams are black as anthracite.

There are no seasons in this flat,
though out of doors the sun shines bright.


### 2

Hell is other people, Sartre thought,
good companions for eternity.

The gaping O of self becomes a nought,
yet Hell is other people, Sartre thought.

We make mistakes however well we're taught
to conjugate the restless verb 'to be'.

The Hell of other people, Sartre thought
gives good companions to eternity.

Heaney, Hughes and Hill and Harrison—
top poets' names begin with H.

A team beyond comparison,
Heaney, Hughes and Hill and Harrison.

They're so grown up, we're in the crèche;
they're generals, we're garrison.

Heaney, Hughes and Hill and Harrison—
top poets' names begin with H.

## 4

The nearly mindless triolet
is the one form that Auden shirked.

With Wilde it turned quite violet,
the nearly mindless triolet.

And drops of Baby Bio let
potted plants seem overworked.

The nearly mindless triolet,
the only form that Auden shirked.

# SUSANNAH AND THE ELDERS

If you knew that this was Mrs Cartwright
just home from Santorini
with a truly super tan, you could be
halfway to justification as a peeping-tom,
the repairs to the building having
exposed the bathroom in a positively
Versalian manner, the light brickdust
talcing her terracotta feet beneath the towel
as she decides it is time to shave her legs again
for the last party of the summer. Such are
the practicalities of living,
skin superbly intense as a location of light,
mirror misted over until one arced swipe
restores its two-way voyeurism
and your disappointment at being separate
less intense than your relief that attractive creatures
are their own dilemma, or presumably
the special traffic of distinct personae
cluttered up with memories of meals and money.

Once they painted things this way
and they insisted on virtuosity
of technique. Researchers ask
how many children did the painter have
or when did he become a Catholic?
What sort of provenance would you provide
the experts with if you told them that
you often encountered Mrs Cartwright
at get-togethers in the Square?
Will the mind as readily as the wrist
limn the surfaces of lust? Tomorrow
we shall pour ourselves into the street
to bring such worries to the barricades.
Tomorrow or next Tuesday. We get only
the poorer Arabs in this quarter,
servants of the ones in darkened cars
and often with glaucoma. What joy
to see the risen flesh as separate
from the soul and independent of
continuous late-night colloquy.

# THE WORLD'S WEDDING

The most dangerous people are those
for whom the present is the only reality.
There is no mystery appropriate to them,
no season of loneliness for disappointment.
Now cut the carousel of slides to show
a round-towered church with pheasants
up to the door, an expedition of content
when all you'll hear is cheeky hopefulness
at this, the noisiest of betrayals—
there are those who'll go down unmade roads
or leap the shuddering pay-train just
to keep their RSVPs true,
hurrying in heat and high-heels to
the prophetical weddings of the world.

Pictures painted stand for ends beyond
ourselves, while photographs are pinned
by truth to be the epilogues of life.
Shun then the snaps that tumble from
the airmail letter (Cousin Circe's wedding)—
look above the chimney-piece—in browns
and dew-flecked mauves a river scene with cows
is half the history of the boarding house
and half the plains of hell. We have to make
accommodation of the separate oracles—
don't go down to the woods today or what
shall we do to be saved? Nobody quarrels
with banks and barristers, but the reckoning
to be frightened of is a dusty scene in oils.

You've followed me so far. Of course you know
it's dreams I'm talking of, of which all pictures
from the quattrocento to the ROI
are nothing more than shorthand reveries.
Thus, when we get the chance, we crowd into
the topless tents of Camberley to sip the fizz
and glimpse the bride changed for the journey south.
Changed she will be, but be less deployed,
her head upon Italian pillows, than when once
she sought a wedding every day, response

47

to being brimfully alive and hating life,
uncoupling what her feet felt from the hope
of arms, making from her mother at the ironing
a massacre of golds and mortal tints.

Speed is eloquence, rushing on to judgement;
Nature the bardic never blots a line.
See how these worriers, neater by a franchise,
settle a grid of surface treachery
on everything; tracing the referent
is basting St Lawrence. We need such nuptials
of world and world if only to catch up
with dismalest relations, glass in hand.
Didn't I meet you in that dream, the one
of tip-truck and tricycle? Who said
the silver-wattled pheasant had to die?
The honeymoon car's a nimbus when sunlight stripes
the headstones. Imagine them now, keepers and carvers,
generations marrying in death.

# DISC HORSE

'Hi, folks, this is the high season for forks,
for getting out the Esky and the Ute,
packing beer and wine and eggs,
us waddling and well-fed urbanites
setting out to trap the sun
where the river picnics on its raft of lilies
and pelicans dock beside the swans.
What horrors our ancestors must have known,
German names, obscure cemeteries,
a hierarchy of the various hopes.'
                    That's it, my voice!
You recognize it now, the tape of language
trying to sound like thought: a riderless phantom,
haunter of a million articles and host
to ego at the conference. Round and round
go galaxies of talk and everyone knows
not to intrude on anyone else's space
for fear that no one then will have the power
to recuperate the narrative.
                    Old friends, new antitheses!
What if instead we go on automatic pilot?
                    Zgtfpxxlrjkdxrhgggkk . . . ooof
                    Sound poetry . . .
'What a work of peace is a man,
how golden in season,
how fond of the infinitive,
in porn and loving how depressed and animal,
in faction at all angles
and apprehensive of the gods,
the futility of the world,
the marathon of cannibals.'
                    Something has gone wrong.
All afternoon we zapped the ether
like Mazeppa strapped to history
with every famous name agog
and buzzing in the Junior Common Room—
'Go down with floppy discs on every side,
                    Inscribe! Inscribe!'
These texts mean nothing, even Shakespeare's,
it's the shape the theory makes that counts.

49

Lost in the dark
an old lyricism keeps coming back
through breathless new lacunae: antiphons
of ageless matriarchy tell the son
he has to pay the price of discourse.
Enter a hot room, past the potted plants
and whispers of Vienna; listen to the Big Bad Wolf
of Parsifal Road—the jockeys of our joylessness
are liberators really. What if we are born
in the colours of compulsion, seeking on the moon
our pale lost wits, sfumato of the sun.
Pack all the artbursts there have been into a box—
they would not tip an inch against
a single phrase of truth. Unless you find
your misery you will rage retardless
up among the fox-furs, furioso of N8.
             No, apocalyptic steed,
canter closer from your turning world
and bring us voices mandevilled
upon the trees of night. After the blaze of music
and the tang of immortality
prepare to listen to those other sounds,
that heartbeat which has given us syntax,
the trope of nerves we say is sweet
erasure, proving that displacement
makes divinity—the god of itself talking
in shape of its most favoured deputy,
a much too loquacious poet
hanging in his participles,
denying everything, including death.

# LAZARUS UPWIND

Despite my news of darkness
ready-made to sell to a Sunday newspaper,
my greatest contribution to humanity
has been embarrassment.

I was in an oven, netted in string,
plugged by garlic and stuffed
with myself, and felt the microwaves,
then split and fell through my own cracks—

As the composer said, I was no longer
a musician, I was music,
and so I wept—I could not hear myself.

Lapis lazuli, the lap of Lazarus,
a gold watch for the creature lasting
and police arriving by a special train . . .

To boldly go where none has gone
and bring back just some episodes
of unction, of the world as onion
floating in the bourgignon of space,
all the sacreds and profanes
concerned with eating . . .

Just a slipping through the door
from an awayday, picking up the mail
and bumping into someone's bike,
then the mumbled words . . .
            'Bexhill was dull
as you'd expect; even the one redeeming
second-hand bookshop shut,
extraordinarily cold for summer . . .'

How can I tell you what you want to know?
You have imagined it beforehand.
It was not like that,
it was not like any of the things I've said it was,
it was like nothing claiming it was like.

You should have moved away
from consequence. One thing I learned,
there are no words within these words,
only a body made for dying,
talking, talking, talking . . .

# THE MELBOURNE GENERAL CEMETERY

This is a territory strange
To me, not the dead's embankment
But this southern city's range
Across a watered sky, the scent
Of burning, gravity's exchange.

My kept-neat city friends, the dead
I'll never meet, seem quite at one
With me. They are, no doubt, well-read,
Their stones and mounds dry in the sun
Following rain. They read in bed

Good news which goes on being news,
Old letters held before their eyes.
What we above must sift for clues
To them is fact and no surmise,
The loss of all there is to lose

Is not a loss at all. They're back
Where heavy Nothing weighs an ounce,
Where Truth is in some distant stack
And all the clawing notions pounce
On shadows in their dawn attack.

Behold amid the left-behinds
A cautious man-amphibian:
Once groping on to shore, he finds
He breathes both elements and can
Take his chance of creeds and kinds.

Almost able now to live
In life or death, this stalker looks
At those he loves, appreciative,
And knows them pictures and the books
Of Thanatation his to give.

# SOUTHSEA BUBBLES

The box tree in the garden
helped navigators guide
ships to New Farm Wharf,
or else Grandfather lied.

He might have done, he had
a beard like Bernard Shaw
and nothing much beside—
that was in '94.

We came from nowhere, a knock
on a wooden citadel,
we had a family tree
sited in hell.

It was Adam's house
our forebears left
and all our felonies
were for his theft.

The past a clear sky
and we the nimbus grouping
but the rain brings
grasshoppers looping.

Thick among the eucalypts,
paspalum and lantana
and Moreton Bay figs
dropping manna.

Boys in serge and parakeets,
alike the sun's darlings,
share watered grass
with strolling starlings.

Prep Schools and private griefs
take sprays and wreaths
to honey fields of death
where hot bees feast.

Not theirs history's bubbles
chained along the beach,
bright convicts of deportment,
each by each.

Days improve, new sons
go sooling a dull nerve,
South Coast Hi-Fi tracks
an azimuth curve.

The honeymoon river runs
past a rotting wharf,
the children of light
are in the surf.

Play history for melody
and not for truth,
possums and sparrows
waltz the roof.

And I must hear before I die
this oracled South
speak true love
from a lying mouth.

# SPIDERWISE

*To Clive James*

## Trapdoor

The origin of metaphor is strange.
As boys we used (but don't let me forget
I only watched, I wasn't very brave)
To put two spiders in a bottle, wave
It over flame, which usually made them fight,
Or flood them from their deep holes for a change.


These were the deadly Trapdoors whose one bite
Sent an inclusive poison racing through
Your veins: I think we thought the risk absolved
Us from all guilt, our cruelty dissolved
In danger. I used my fear of football to
Ward off death fears in the dorm at night.


And then I thought of dying on the field
When someone passed the ball to me. I say,
When challenged to declare what virtue I'd
Like most to have, 'It's too late now to hide
From balls and tackles, you can't get away;
If I'd had courage, I'd at least have squealed.'


We are shut out of our own universe.
Perhaps the spiders in the bottle howled
Or cursed those lumpen schoolboy gods of theirs
Or practised spider jests upon the stairs
Where tyrants crack because one wit has scowled.
We fume and Rupert Murdoch's none the worse.


And writing in this corset-stanza may
Be nothing more than flying in the face
Of new technology. And now they're topping
Up the latest tank of history at Wapping—
A metaphor is when you have one space
To fill and all of life to file away.

Most metaphor, as Kipling guessed when he
Made Shakespeare witness as a child the end
Of kittens at a sister's hand, is home-
produced. The tribal creature is alone
With only tribal words to help him blend
His uncloned self with all humanity.

But metaphor is often out of date.
I find a powerful trope: 'the straitjacket
Of all our childhoods', but I've never seen
A straitjacket—drugged, behind a screen,
The madman of today can raise a racket
And none of it will reach the ears of state.

We live on dead skin of the mind. To write
Is to commit oneself to the past tense
Within the present act of mixing words
Whose immanences settle as occurreds:
And so we pay with pence these bills in cents
And stoke a microchip with anthracite.

The language will not move except to laugh
At Big and Little Enders, Old v. New—
Thus pedantry is publishing a stricture
While TV Gnomes are 'massaging a picture',
The dictionary hasn't got a clue,
The wind blows out the door both wheat and chaff.

The long relays of childhood can provide a
Range of wrongs to stake a moral claim.
The lie of words (O ambiguity!),
The rust of rhyme, short-cut of simile,
Make yours and mine and Swift's desire the same—
The angry soul is quite a bottled spider.


*Redback*

My next portmanteau-spider is a cert
To raise Australian hackles (and bare bums
Off dunny seats): the Redback scuttles out
From underneath your fallen strides to scout
The scene. You always wonder what becomes
Of him when brooms and Lysol lay the dirt.

But let the Redback be my image of
The naturally malevolent activist,
The dried-blood cross upon his back be rubric
Of the Unsettled Terrorist, his Kubrick
Horror-sequence of a bite de-mist
The mirror of normality at Sydney Cove.

Unlike the Trapdoor he is quite at home
In human homes. Communications architecture
Suits his way of feeding on the grid
Of everyday existence. He stays hid
In myriad corners, happily the vector
Of challenges from brushed velour to chrome.

I'd like to think that he was waiting there
When that initial cargo of strong rum
And human refuse hit Australia's shore.
To flog, to dream, to endlessly explore
Might make Arcadia of a rural slum:
The Redback gave the hedonists a scare.

'Simply the thing I am shall make me live.'
Bravo, the Redback! But he stands for more.
That some shall live, some others have to die,
What my will urges, your will must supply—
The Crowded Ark's the spider's metaphor:
He has to eat, and then he will forgive.

And he survives the change of gear which
Pulls Australia from remote back number
To Brave New World of Opportunity.
His atavism's there for all to see,
Devoid of any Panglossian lumber,
A backyard warning to the newly rich.

Especially as we've tamed the whole wide land
To seem Pelagian peninsula
Where terrors have a user-friendly feel.
The bush looks Art Nouveau, the wattles steal
Their hazy shapes from Mucha; in the car
The Esky waits the barbecue's command.

No doubt the mainframe's large enough to cope
With sharks and bushfires, droughts and mining rights.
The people have been told they're legatees
Of freedom lovers from five centuries:
They've made stigmata of their appetites
And trust in God and in their horoscope.

The Redback understands. He's his own priest
And has an Opus Dei in each leg
To race him to the necessary prey.
Perhaps he likes to keep Australia Day
Among the stacks of stubbies and the keg,
And wave the Southern Cross above the feast.

He has no need of words, a true numbskull.
He will evolve no further, like the shark,
But at cohabitation is a whizz.
How subtle this philosophy of his,
To warn us, as we fumble in the dark.
Relieving Nature won't be comfortable.

# THE DULL MAY WAKEN TO A
# HUMMING-BIRD

Now it is neither dark nor light,
swimming off Liverpool, coming ashore
on a shingle beach to valiant figures saying
'Entertain me, take the printed wrapper
off conjecture.' Surprising it is not surprising
later in those pea-fields with fast friends
along a brushwood track, then to watch
a flower exploding like a sneeze, 'something
to do with its way of reproduction'
says the voice of explanation,
reductive but obliging, and beside the flower
a bird which closes up into itself,
sublimest sphincter acting out a joke
but nicer than that paid-up omphalos
where time is tucked away. A Royal
Commission is sitting all the daylight hours
investigating the word 'when', a hundred lawyers
lounging in the court and all the parts-of-speech
on call: Justice is known to be a case
of polyester versus cotton and no decision
yet expected. Three friends are watching
as a four-engined jet comes in
for emergency landing on a playing field.
From the perfect crash we pull out screamers
as the fire spreads along a wing, each one
inside his harness, and still the dead
are thought to be spectators at the last
gymkhana of the season. Threatened by burns
which you could feel through sleep, creeping
up behind the murdered rainmaker, a murmur
giving warning of resolve and all the feet
of relatives piled in the poring dark,
the self is hailed—'I know you, you are he
whose shadow fills the hemisphere,
a vision twice as wide as chaos.'
A second voice is singing in the leaves,
vessel of the insubstantiality of love:
'You are none other than that unpronounceable
Bohemian whose quartet was adapted
by Franz Schubert; you are a tone of talent

lost in the splendour of the universe.'
Greyness spreads like sand through fingers
as they untie the boat beneath the little wharf
and leave this coast of fearful breakfasting.

Interior waking that the moon draws up
or slavish constancy of blood
breeds thus the contiguity of words
and feeling. Tell the sons of Freud
there are no templates and no temples,
only the ancient harness of grey thought
which dresses us for true extinction,
merely the wished-for continuity
of age and enzyme. Remarkable under the sun
this everything of memory,
the first goodbye of coming to the world,
the final entry in the book of gold.

# THE STORY OF U

And now the track is snowed by words,
The poor train of childhood followed,
A good aunt picking out the thirds
On an old piano, gutted, hollowed
By years which left the trees the same,
Adding one storey to the house
In others' hands; and can you claim
That here sex showed you her old powers?

The little ghosts which charmingly
In gentle masochism shone
Grew up and lived oppressively
Till loving was a looking-on;
The staff was joined by Feminists
With good French accents and strong wrists,
Then blinds were drawn and hands went free—
'Une dentelle s'abolit'—

Flesh at its most jardinière
Will always be the watcher's prize;
Why then do anymore than stare,
The sonnet tells you that her eyes
Are only words and time is time?
And even on Banana Downs
The Fetishist of Polished Rhyme
Will spill his soul in abstract nouns.

This is the house they made for you
With water-steps and angled palms,
A cellar where your tears came true
And terrors took you in their arms:
Down by the water a boatshed
Collected the dynastic dead
Who heard cicadas keeping on
Their etching of a single song.

The torturers have moved upstairs.
Elegant explainers leave
Missals in their chests-of-drawers,
The labyrinth of just one sleeve
Now takes you through to violence
And precious parities of women
In Mallarmean recompense
Prove the pain and rage are human.


## THE EMPEROR HADRIAN

*Animula vagula blandula . . .*

Little soul, like a cloud, like a feather,
my body's small ghost and companion,
where now must you go, to what region?
Pale little, cold little, naked little soul
who will you play with, what will you laugh at now?

# THE PANTOUM OF THE OPERA

Life has thrown its acid in his face
and so he haunts the decks and tiers of light.
Susannahs he has watched cannot replace
the mother-love which ran from him in fright.

The booking clerk observed him: yes, he thought,
life has thrown its acid in his face,
and he'll recall each amoroso note
Susannahs he has watched cannot replace.

Tonight the trucker from Emilia sings.
The booking clerk observed him: yes, he thought,
ten years from now heart failure in the wings
and he'll recall each amoroso note.

They've queued for this, they've paid three times the rate.
Tonight the trucker from Emilia sings.
Domingo-fanciers in the slips debate
ten years from now heart failure in the wings.

Sometimes a new-found friend will take him home.
They've queued for this, they've paid three times the rate.
You might get killed, you might just wake alone,
Domingo-fanciers in the slips debate.

Swart Papageno practises his bells,
sometimes a new-found friend will take him home,
and Don Giovanni has a choice of hells:
you might get killed, you might just wake alone.

Our phantom selves dance wildly when they hear
swart Papageno practising his bells.
Say No to Heavenly Takeaway; it's clear
that Don Giovanni has a choice of hells.

Say No to Heavenly Takeaway; it's clear
Susannahs he has watched cannot replace
the mother-love that fled from him in fear.
Life has thrown its acid in his face.

# DOES A RAKE GO TO A BROTHEL TO SING?

*(In memory of the creators of* The Rake's Progress)

We have been deceived by our idealists—
Tom Rakewell acts the audible
and not the consequent: for something to be real
it must be possible to sing it.

And we can sing the starting of the world,
a balancing of love, the games of touching teeth,
the desert dreams of conquerors,
yet wake beside the innocentest teacher
in real time, kept shadowless
beneath the cuckoo clock's retard.

Dreams are the grandest operas,
unruined by a Gounod or a Meyerbeer.
They cannot be cured with meaning
but must sing the very tones of happening.
So tell our father we are blood and soul for him,
we are plainly set in place
as blades of grass, and should we die for love
it will be love of syntax. Who are these
punk phantoms of Pontormo? Who sits fat
in Heaven, looking lovable?

Judgement is all Creation sings.
Here we go back to finding crimes
to match the punishment. Our needs
are music, water, persiflage,
a set of values on a colour card.
No wonder then our rulers subsidize
an art you are expected to dress up for.

To dress for dreams is dressing up forever.
Mother Goose has loosed her stays
and let her hair in delta flood
a veteran champaign. It is too late always
if you're lucky. A-Major sounds within the ranks.
*Sweet dreams, my Master. Dreams may lie,*
*But dream. For when you wake you die.*

# ON TOUR

I had just finished the first part
of my recital. Throughout I'd noticed
a small quiet man in the front row
wearing decorations. An Austrian official,
shunned by the Milanese,
but sure to speak impeccable Italian.

In the Interval, he'd have liked to speak to me,
but more importunate faces intervened.
'We look to the North, you know,
we are Italian, we are Catholic,
but we are not backward facing.
In the North lies all our promise,
we hope you will conduct Bach's Passions
here for us when next you are on tour.'

In the second half I played
my 'Variations sérieuses'
but felt that Beethoven, Weber even,
might be more the sort of thing they'd like to hear.
Accordingly I broke my programme
and finished with the *Waldstein.*
I don't think I've ever made the rondo go so well.

Amid the bravos and the knowledge
of all the toasts to come, I settled
my Protestant soul by saying to myself,
'You aren't a real Bach or Shakespeare
but a Jewish convert with a flair
for counterpoint.' The little man was there
beside me, humble but intrusive—
'You must forgive me, we do not often meet
such talents here in Italy.
I am Austrian but prefer this country
to my own. I am not without distinction
but I loathe Vienna. Mozart, at your service,
Karl, the elder son. Won't you please
play something by my father.'

# THE LOUD BASSOON

Out of the sound swamp, the delta of dreaming,
Shuttles reluctantly what is accountable,
Warp-words, intrusions, rubbishy mutterings,
Harvest of nicknames, hierarchical slop,
Announcing like radios always left on
Through thin walls of flats in scaffolded towers,
Infusions of wonderment, brought to us daily—
X's great symphony, *The Inconsolable*,
His Number Four, or that tragic opera
*Gustav's Vasectomy*, a cycle of poems,
Sadder than cypress, *Silicon Mandibles*.
Always a surf of creation to hear through,
Tinnitus feasting on blood and on sunlight
Bringing a past which is parody-present,
Ululitremulant, warning that Nature
Mucks out impatiently, scatters what flesh thinks
Echo-eternal. The ear's on an island
Centuries heavy, lost in geography,
Swum to by marvellous patterns and patents,
Though Self will not find it. Orchestral voices
Beam in so seriously abstraction sounds
Like Swinburnian hendecasyllabics,
But duty sidles up to each wedding guest
And its interruption in story or fact
Spells out the message in mendicant latin—
*Mors aurem vellens, vivite, ait, venio*
(Death plucks my ear and says Live! for I come).
We need translations who will be translated,
High blessings bestowed by bloodstream and logic,
Trees to sit under and hear the sky stutter
Arcadian instances robed as statistics.
Now morning unsettles the dust in shut rooms,
Airwaves revive with the titles of living,
*The Wrong of the Earth*, old words in new harmony,
Listened to lovingly, one cat on your lap
And one in a sun-stripe—how can you bear it,
The clatter your heart makes as it challenges
Air and the universality of air.

# PONTORMO'S SISTER

The world's face is a woman's
who died early, the smoothnesses of life
waiting on a ruled horizon—
Consider this in profile even
in my *Visitation*, the four ages of Woman
unable, like God, to be anything but themselves,
and in Mary it seems lit within,
myself there, swept by darkness.
All my people are the same person,
as every artist shows: there grows a face
as hedgerows grow, as water shapes
in droplets when it falls, as we emerge
from the doors of dreams to be ourselves.
Piero's faces never vary, did he perhaps
have a sister who died too young to marry?
That's how I found technique,
a way of bending Nature to the line
of my depression. We say at twenty
and at forty and at sixty, there are
measures and distinctions you call art—
but no, we're in the shambles
with our little sisters and our parents,
we're tied to flesh and death forever
while we live, and out of it our masters ask
'Make me Veronica, the dogs and boys
grape-picking, Jesus faltering beneath
the cross's weight.' I can paint a word
if the word is death, but what I cannot do
is show it to you
unless I wrap it in a nimbus.

O little sister dressed in death,
I have painted you in everyone
and now I beg you draw the veil
across my eyes. It is time for me
to sketch God's face, a smudge of grease
on old familiarity. This is the message
of the mannered style: God looks like
anyone who ever lived, but more so.

# CROSSING THE TIBER ISLAND

This is God's Circus Maximus—
a fledgling sparrow slides from pavement to gutter
and miraculously avoiding Rome's traffic
lives to skid by panic wing-power
into the opposite gutter
and crouches there dynastically.
This is no true contest,
Rome has turned its thumbs down
on yet another creature, the Tiber Island
shakes with heat and gravity.
Are not two sparrows sold for a farthing?
They would not fetch so much today
in this expensive city. Gods come here to die
and now a grist of shit and faith
covers Romulus's mound to seventy feet.
We cup our blood for dreams to drink,
thinking we have so many good and evil acts
to chorus us, stones of faith outstaring nerves
beside trompe l'œil ceilings where
bath night fronts a deathless Pantheon.
Each human mind is Rome ruled by a mad
and metaphorical Emperor, or Pope
praised for piddling fountains
and passing barley-sugar baldachinos.
Stepping from the hotel's air-conditioning
you become a city sight-seeing in a city,
your history is just as marvellous as Rome's,
your catacombs the haunt of pilgrims
from Hyperborean archipelagos,
new worlds pillaged of their optimism
to gild a lavishly despairing faith,
the Tiber Island is the food you eat,
the cows ten times your weight, the little
prawns inside their carapace, and lettuce
dressed to ease Christ's vegetarian pain.
The sparrow in the dust
knows neither Pontifex nor Aesculapius
but twitches on its bed of wings
terrorizing Heaven and whichever
deity could meet its dying eye.

# PISA OSCURA

You know how images keep coming back,
The lifted arm before the heart attack,
Yet out of all the basket-work of shapes
And plots, those vandalized electroscapes
Of daytime dreaming, how remarkable
The least significant of them is able
To light the mind and flood the memory!
Don't introspect if you want honesty,
And that's what Freudians presumably
Intend when fixing eyes upon a past
That's like a slow vertiginous open-cast
Whose work load is regrettably colossal,
Its every truth impacted like a fossil.
So holidays from thinking look like cards
Of saints in shrouds and girls in leotards
Proclaiming less a haunting charm of face
Than unexpected valency of place,
Or so I felt, midway from sink to freezer,
When there before me hovered dusty Pisa.
The town's historical, a saucer round a cup
Of regional accidie, a dried-up
Vacuumed-out Ligurian emptiness—
I've been there often and enjoyed it less
Each time until this year when suddenly
Of all the history-pitted bits of Italy
I found it most like home, a proper cage
(Not Pound's) to hold the ageing spirit's rage.
Streets almost empty, traffic ice-cream slow,
The silent squares out of de Chirico,
All tourists heading for the Leaning Tower
(The bars have yet to learn of Happy Hour)
And history's ghosts so sullen they won't come
Into the present at the Guide Book's drum.
Of the famed Piazza dei Miracoli
This poem has nothing much to say—degree
Of leaning from the vertical perhaps,
The bombs that made Benozzo's frescoes maps,
Incorrigible youth which carves its name
On G. Pisano's pulpit without shame,
Perhaps to discount in one heresy
The intolerable weight of history

And catch the tired tourist as he gapes
In wonder at the time-defying shapes
And with a CARLO or an ELVIRA
Restore the present in a vandal's scar.
The private miracle the site enshrines
Is in the meeting-up of marble lines,
Geometry inscribed on empty sky,
Invisible gods held fast by symmetry,
And all the dead, great figures in their day,
Not knowing that their names have worn away,
Insisting still in pompous silence that
A Campo Santo is no Ganges ghat
But, filled with dust brought from the Holy Land
And peregrine of faith in sacred sand,
Retains a monumental gesture for
The rotting body and redundant law.
The rest of Pisa sleeps beyond this square,
Few tourists break their scheduled journey there
To wander back towards the river and,
Cascading maps and ice-cream cones in hand,
Quiz Ugolino's tower and Shelley's garden
Or Byron's palace—value judgements harden
When dull façades and husks of buildings lour
On a weedy yellow river and each door
Is fortified by bars and rusting bolts
And echoes only to the Fiats and Colts
And Renaults charioteering the Lungarno,
Italia Martire ma *cum grano*.
Yet here, before a bridge, tucked in between a
Lorry and the sky, the della Spina
Church appears, God's jewel-box, a toy
Created for that icon'd marvellous boy
Italians in their hearts have made of Christ
(His rape of Heaven, His Redeemer's heist!)
To shrine the thorn which tore His silver skin
Before the clause of Godhead thundered in.
This Lilliputian church, not Dante's spite
Brings Pisa from an untransfigured night
Straight to my dreams—this is the grace of love
That Dante's terza rima cannot prove,
The reconciling shape which frees mankind
From murderous faction of a poet's mind.
If medieval wholeness ever was,
This is its only symbol, its True Cross,

This, while the turning wheel of faith revolved,
Showed saint and sinner all would be resolved,
That, with his fondness for 'I told you so',
Dante was just some foiled Castruccio.
A similar anathema still sends a
Shudder through me in the Sapienza—
No university could survive that name
And modern seats of learning trim the flame
To safe and low accountability,
Their Galileos home in time for tea;
Though one emboldened tutor broke the spell—
'Let me be Virgil, I will show you hell!'
For hell, as Shelley said, might be a city
Much like London, dressed in cold self-pity
Fanning-out in grids from dread of death,
Its towers of hate above, its sewers beneath
Where flows the dreck of self—the squares, the prisons,
All at the service of destructive visions.
This in my chill mid-morning I recalled
As Autumn vapour pecked and spilled and stalled
Around my window; a city of the mind
Whose used-up living lives on in its rind,
A Pisa worse than the exhausted South
Despairing ever at the river's mouth,
A shadow city, formed of self and soul,
Its past pristine, its present on the dole.